Making Pistons for Experimental and Restoration Engines

Stephen D. Chastain

B.Sc. Mechanical Engineering and Materials Science
University of Central Florida

Making Pistons for Experimental and Restoration Engines

By Stephen D. Chastain

Printed in USA

ISBN 0-9702203-4-0

The Small Foundry Series by Stephen Chastain
As of 2004:

Volume I. Iron Melting Cupola Furnaces for the Small Foundry
Volume II. Build an Oil-Fired Tilting Furnace
Volume III. Metal Casting: A Sand Casting Manual Vol. I
Volume IV. Metal Casting: A Sand Casting Manual Vol. II
Volume V Making Pistons for Experimental and Restoration Engines

stevechastain@hotmail.com
Steve Chastain
2925 Mandarin Meadows Dr.
Jacksonville, FL 32223

WARNING – DISCLAMER

This book is to provide information on the methods the author used to make replacement parts in a home foundry and machine shop. Both machine tools and foundry work can be dangerous. No attempt has been made to point out all of the dangers or even a majority of them. Although the information has been researched and believed to be accurate, no liability is assumed for the use of the information contained in this book. If you do not wish to be bound by the above, you may return the book for a full refund.

Warning: Molten metal and high intensity combustion can be dangerous. Incomplete combustion produces carbon monoxide, a poisonous gas. Only operate a furnace outdoors. Stay clear of all ports when a furnace is in operation. Observe all rules regarding safe foundry practice. Do not attempt to melt metal if you are not qualified. Do not use gasoline or other low flashpoint fuels to light a furnace. Do not spill molten metal on yourself, others or any wet or damp surface. Always wear protective gear. Observe all regulations regarding the safe handling of gaseous and liquid fuels. Safety is your primary responsibility.

TABLE OF CONTENTS

PURPOSE: The purpose of this book is to provide simple manufacturing solutions for the production of workable parts for restoration or experimental internal combustion engines. While these processes may be too time consuming for a large commercial venture, they work well for short run and small-scale production.

Because this is book 5 of the *Small Foundry Series,* it is assumed that the reader, by now, is at least familiar with the sand casting process. Only casting topics specific to the piston project will be discussed. The reader is referred to *Metal Casting: A Sand casting Manual for the Small Foundry Vols. 1 & 2* for general casting practice.

It is assumed that the reader has some machine tool skills and is at least able to make the most basic cuts on a lathe and a vertical mill. Some of the descriptions may appear too basic for the experienced machinist, however they would be helpful to the novice, therefore they are included.

Modern design and analysis are done by modeling the piston on a computer. Pistons have been around much longer than computers; therefore some of the older material regarding piston design is included. The results may or may not coincide with modern methods, however it is introduced to provide a background pertinent to the era in which the parts were produced.

Cutting Pin Retaining-Clip Groove Using a Shop-made V-block Vise

INTRODUCTION: Old engines have always fascinated me. Several years ago, I discovered an antique 4 cylinder flathead half buried off a riverbank. It looked pretty bad but being a novice, I assumed it was probably discarded because of carburetor or electrical problems, making it an easy fix. After getting the OK to remove it, I hauled it home, the whole event becoming the source of amusement to many. I soon discovered that the engine had been full of water for years and was completely frozen up. Many parts crumbled to dust upon disassembly. Replacement parts were virtually nonexistent, and those that were available cost several times what a working machine was worth.

I had recently purchased a 12 x 36-inch lathe and had managed to learn a few basic cuts. At this point, I had nothing to loose and everything to learn, so I set out to make all of the engine parts myself. I discovered that parts were fairly easy to produce. Soon, I had all of the parts made and the engine assembled. After a little electrical trouble shooting, the engine came to life. It fired up almost immediately upon touching the starter switch and ran with a health roar! The engine ran and it ran well. Soon all those who doubted were saying "we knew you could do it." Since then, the engine has powered a 10kW backup generator and accumulated hundreds of hours of use.

Over the years, I have taken on several other restoration projects, many referred by the local technical school. Each has been a rewarding experience. The point of all of this is that: *home made parts work and work well!* Lack of parts is no longer an issue when you can make them yourself. Blocks that have been bored oversize can be cleaned up and fitted with custom pistons and modern rings. Those impossible projects become viable when you can make your own replacement parts. You and your friends will be surprised at what you can do with a small foundry and a few machine tools. I currently drive a 1932 REO car with homemade pistons and bearings, among other things.

BASIC DATA AND PISTON DESIGN:

The simple looking piston performs many functions. It must transmit the force of combustion to the wrist pin, transmit the absorbed heat of combustion to the cylinder walls and hold the piston rings so that they may effectively seal the cylinder.

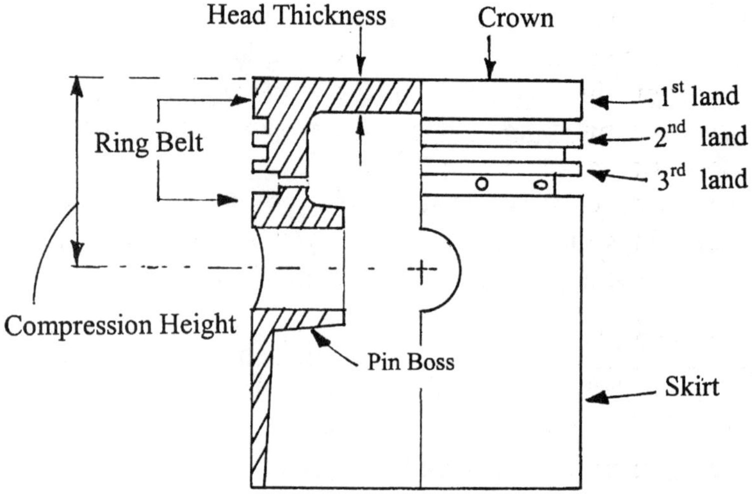

Piston

The main parts of a piston are the top, which may also be called the head or crown, the ring belt, the pin bosses and the skirt. The top is part of the combustion chamber.

The top may be flat, or a combustion chamber may be cut into the top of the piston. The top may be raised or have a bowl cut into it. Soot contamination of the lubricating oil in diesel engines is reduced when the combustion chamber is located in the piston, as opposed to the cylinder head.

The ring belt usually has three or more rings. Two cycle engines do not require oil rings and therefore may have only two rings. Ring lands are located between the ring grooves. The top land, or first land is located above the first ring. The second land is heavy because it supports the first ring and bears the majority of the pressure and thermal loading of the ring belt. The second and third lands are lightly loaded. Because of expansion of the piston top at operating temperature, the ring lands are usually relieved or cut smaller in diameter than the rest of the piston.

The pin boss supports the piston pin and transmits the force of combustion to the pin. It is one of the most highly loaded areas of the piston.

The piston skirt, which wraps around the lower part of the piston, distributes the side loads and prevents the piston from rocking in the cylinder. Long pistons rock less than short ones and are used in diesel engines to reduce the number of required compression rings. It is common to see 2 gas rings on pistons of 1.4 bore but 3 may be required when the length is 1.0 to 1.2 times the bore.

Common Dimensions of Modern Aluminum Pistons

(Relative to Diameter)	Gasoline Engines		Diesel Engines
	Two Stroke	Four Stroke	Four Stroke
Diameter in inches	1.375 to 3.0	2.5 to 4.25	3.0 to 7.0
Length	0.8 - 1.1	0.7 - 1.0	.9 - 1.4
First Land	0.06 - 0.10	0.06 - 0.12	0.10 - 0.20
Second Land	0.04 - 0.05	0.04 - 0.05	0.07 - 0.09
Compression Height	0.40 - 0.70	0.35 - 0.60	0.5 - 1.00
Pin Diameter	0.22 - 0.30	0.25 - 0.30	0.3 0.44
Pin Boss Gap	0.25 - 0.40	0.25 - 0.40	0.3 - 0.46
Head Thickness	0.07 - 0.10	0.07 - 0.10	0.10 - 0.20

High mechanical loads are usually restricted to the support of the top ring and the pin bosses. The first ring groove is highly loaded both mechanically and thermally and is of particular importance. Several factor influence the temperature of the first ring groove and are summarized here. A speed change of 100-rpm changes the temperature of the first groove by 4° to 7° F. Variation of the ignition point by 1 crank degree causes a temperature change of 2° to 4° F. Raising the compression ratio by 1 unit causes a temperature increase of 7° to 22° F. However, because of increased expansion of the charge, the exhaust gas and cylinder head are cooler. A load increase of 14.7 psi, at constant speed, increases the temperature of the first ring groove about 18° F.

Thermal loads are often larger than mechanical loads and may dictate the design. Thermal loads can be calculated in pounds of fuel burned per square inch of piston head area or in (brake horsepower) bhp / $in^2_{\text{piston head}}$. Due to aluminum's higher thermal conductivity, aluminum pistons run cooler than cast iron and have a higher output per square inch of head area, when used without special cooling.

General thermal loading for pistons*:

Aluminum	up to 1.5 bhp/ $in^2_{\text{piston head area}}$
Aluminum (oil cooled)	3 bhp/ $in^2_{\text{piston head area}}$
Cast iron	.7 bhp/ $in^2_{\text{piston head area}}$
Cast iron (oil cooled)	3 bhp/ $in^2_{\text{piston head area}}$

*Note that Honda produces racing engines that generate over 4.3 bhp/in^3 at 25,000 rpm

The output of many engines falls below 1.5 bhp/ in^2, 1.5 bhp/in^2 considered the upper limit when using uncooled classical trunk pistons. The carbonization temperature of the oil and the softening point of aluminum establish this upper limit. Modern HD oils allow the temperature of the top ring groove to reach 400°F and intermittently 500°F under full power. Aluminum has good low temperature strength but looses about 50% of its strength above 600°F.

Aluminum's abrasion resistance is also low at high temperatures.

Typical Output per Square Inch of Piston Head Area:

Bore	hp	Cylinders	hp / cylinder	Head Area	hp / sq-inch	Engine Type
2.23	23.4	4	5.85	3.9	1.5	Ford Anglia, 1939
2.3125	3	1	3	4.2	0.714	Industrial Air Cooled
2.625	3.5	1	3.5	5.41	0.647	Industrial Air Cooled
3.0625	65	8	8.125	7.36	1.1	Ford V-8, 1932
3.125	8	1	8	7.66	1.04	Industrial Air Cooled
3.125	85	8	10.625	7.66	1.38	Ford V-8, 1935
3.125	105	12	8.75	7.66	1.14	Lincoln V-12, 1935
3.25	105	8	13.125	8.3	1.58	Chevy V-8 1935
3.4375	175	12	14.58	9.28	1.57	Packard V-12
3.5	32	4	8	9.62	0.83	Wisconsin V-4
3.78	38	4	9.5	11.22	0.846	Ford LSG-423

Because each engine operates under different circumstances, these are general rules regarding thermal loading. Air-cooled engines run hotter than water-cooled. Two stroke pistons run hotter than four stroke pistons because the piston is used as a valve, and the temperature of the cylinder around the exhaust port is higher. Because cast iron has low thermal conductivity, iron pistons run hotter than similar aluminum pistons. The temperature at the center of a cast iron piston head will be approximately 800°F, while the center temperature of an aluminum piston is approximately 500°F.

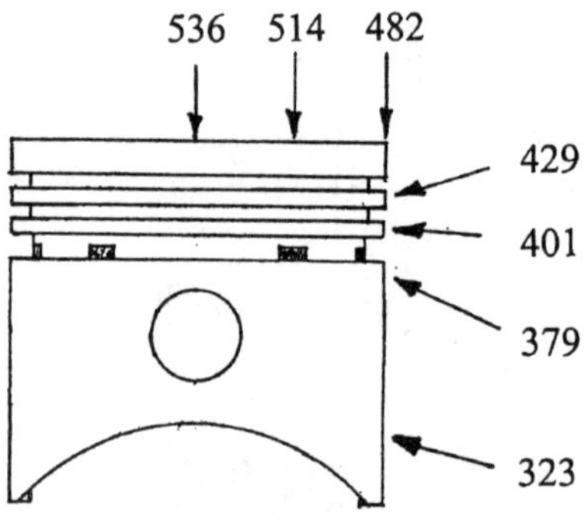

Approximate Temperature Distribution in °F for a 4 cylinder 152 in^3 Spark Ignition Engine @ 4600 RPM Wide Open Throttle

Several methods are used to determine the piston head thickness. Cast iron pistons are almost always oil cooled. The metal sections are made as thin a possible, the actual thickness determined by mechanical loadings. Aluminum alloys have high thermal conductivity and may be used without cooling up to 1.5bhp/in^2 They are designed with thicker sections to conduct the heat to ring belt and skirt. The piston will probably determine the output of air cooled engines. Pistons will be limited to considerably less than 1.5 bhp/in^2 and be made of aluminum alloy.

DETERMINING HEAD THICKNESS:

The head may be treated as a flat plate with a uniform load and rigidly supported at the outer edge.

Thickness of head = $\sqrt{3pD^2/16s}$ inches

p = pressure, psi

D = cylinder diameter, inches

s = permissible stress in tension, psi

Heat flow through the piston head to the cylinder walls may determine the head thickness.

Head thickness for heat flow:

Thickness of head = $\mathbf{H / (12.56c(T_c - T_e))}$

H= heat flowing through head in Btu per hour

c = heat conduction coefficient, Btu per in^2 per inch per °F
7.7 for aluminum, 2.2 for cast iron

$\mathbf{T_c}$ = Temperature at the center of the head, 800°F for cast iron and 500°F for aluminum.

$\mathbf{T_e}$ =Temperature at the edge of the head.

$\mathbf{(T_c - T_e)}$ for cast iron is approximately 400°F

$\mathbf{(T_c - T_e)}$ for aluminum is approximately 130°F

H, the heat flowing through the piston head may be estimated by the formula:

$$\mathbf{H = K}\boldsymbol{Cw}\ \mathbf{x\ bhp}$$

K = the part of heat input that is absorbed by the piston.
This ranges from 4 to 5.25%.

C = the higher heating value of the fuel used

w = the weight of fuel used in pounds per bhp/hour

bhp = brake horse power per engine cylinder

Properties of Fuels: *Higher and Lower Heating Values

Fuel	Specific gravity	Weight per gallon	Btu/pound*	
Gasoline:	.702	5.86 pounds	20,460	19,020
Gasoline	.739	6.16 pounds	20,260	18,900
Kerosene	.825	6.88 pounds	19,750	18,510
Light Diesel	.876	7.30 pounds	19,240	18,250
Medium Diesel	.920	7.67 pounds	19,110	18,000

1 brake horse power per hour = 2545 Btu

Estimating H from brake horsepower per cylinder:

Analysis of fuel consumption per bhp/hour for several gasoline engines gave efficiencies from 22.4% to 27.1% with the average being 24.8%

Assuming 24.8% efficiency, the heat input per bhp/hour is:

2545 Btu/.248 = *10,262 Btu per brake horse power hour*

Example: Arbitrarily selecting the 1932 Ford V-8 at 8.125 bhp per cylinder, determine the piston head thickness:

Heat input per cylinder $_{32Ford}$ = 8.125 bhp x 10,262 Btu/bhp hour

Heat input per cylinder $_{32Ford}$ = 83,379 Btu/hour

H= K*Cw* x bhp

K=.05, (*Cw* x bhp) = 83,379 Btu / hour

H= .05 x 83,379 Btu / hour = 4169 Btu / hour

H= 4169 Btu / hour

Estimating the piston head thickness:

Head Thickness $= H / (12.56c(T_c - T_e))$ inch

c aluminum = 7.7 , $T_c - T_e$ aluminum = 130

Head Thickness = 4169 / (12.56 x 7.7 x 130) = .332 inch

The head thickness is .332 inch, which sounds reasonable for this engine

Empirical formulas are commonly used in the design of automotive pistons.

Thickness of head = .032D + .06 inch (permanent mold castings) (use a safety factor of 1.5 to 2 for sand castings)

Thickness of the wall under the rings = thickness of head
(Because the same amount of heat is flowing through the ring belt)

Length of piston = D to 1.5 D

RINGS AND WRIST PINS:

Ring Belt: About 70% of the heat absorbed by the piston flows out through the ring belt. The top ring land, being close to the combustion chamber, has the highest temperature. Rapid carbonization of the lubricating oil, at about 410° F for non-detergent oils and 485° F for detergent oils, causes sticking of the rings. In order to reduce the temperature of the upper ring, it is placed down from the top of the piston head. Gasoline engines place it between .06 bore diameter to .12 bore. Diesel engines may place the ring .2 bore to .3 bore down from the top.

The second land supports the first ring, which is subjected to the full gas pressure. The second land should be at least equal to the radial thickness of the ring so that it forms a square section. Values of 1.5 to 1.7, the radial thickness, are also used. The remaining lands are subjected to much less pressure and may be as small as .0312 bore, as required to minimize the piston length.

Consulting the ring manufacturers rarely produces reliable ring groove depth information. Simple formulas to estimate groove depth are:

Compression ring groove depth:

$$\text{Depth}_{\text{compression ring groove}} = (\text{ring radial thickness} + .003\text{bore} + .010)$$

Oil ring groove depth

$$\text{Depth}_{\text{oil ring groove}} = (\text{ring radial thickness} + .003\text{bore} + .030)$$

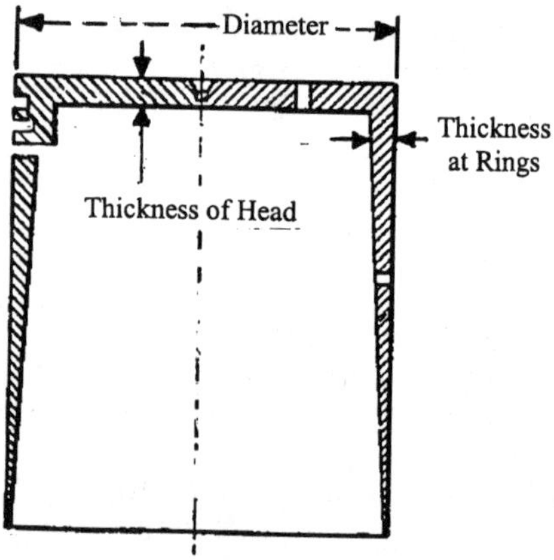

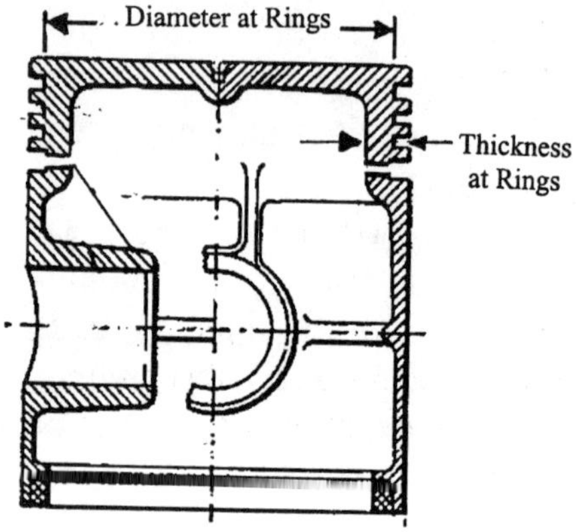

The piston wall thickness, for ideal heat transfer, should taper from the head thickness at the top to zero at the open end. The thickness behind the ring section should be equal to the thickness of the head because the same amount of heat is flowing. A large fillets is used at the inside top edge.

Left: The upper drawing, is laid out for heat transfer. The lower drawing is modified as required for mechanical loading.

EXPANSION OF THE RING BELT AT OPERATING TEMPERATURE: Metals expand with an increase in temperature. The expansion is calculated by using the coefficient of expansion. Each metal or alloy expands at a different rate and has a different coefficient of expansion. Aluminum silicon alloys have a lower coefficient of expansion than aluminum copper alloys. Cast iron has a lower coefficient of expansion than all aluminum alloys. Expansion is calculated by:

Expansion $= K\,l(T_2 - T_1)$

K=coefficient of expansion, l = length, **T** = temperature

Coefficients of Expansion per °F

Iron	0.0000074		
Aluminum Alloys:			
#242	0.0000131	#332	0.0000116
#319	0.0000134	#333	0.0000126

Example: Determine the clearance required for the top land of 3.75-inch diameter aluminum piston of alloy #242 if the piston head is at 500° F and the cast iron cylinder wall is at 200°F. The piston is machined at 70° F.

Piston Expansion = .0000131 (3.75) (500°-70°) = .0211-inch

Cylinder Expansion = .0000074 (3.75) (200° – 70°) = .0036-inch

Assuming a few thousandths of an inch for a running fit, .021-inch is the minimum amount of relief for top land of this piston. I would remove an additional few thousandths as a safety factor for extreme conditions (hot days and heavy loads).

PIN BOSSES: Piston pins are made of 1020 or similar low carbon steel. They are case hardened to approximately Rockwell C 60 and ground to a satin finish.

The diameter of the piston pin is determined by allowing a maximum bearing pressure of 2500 psi. The bearing area is considered to be the

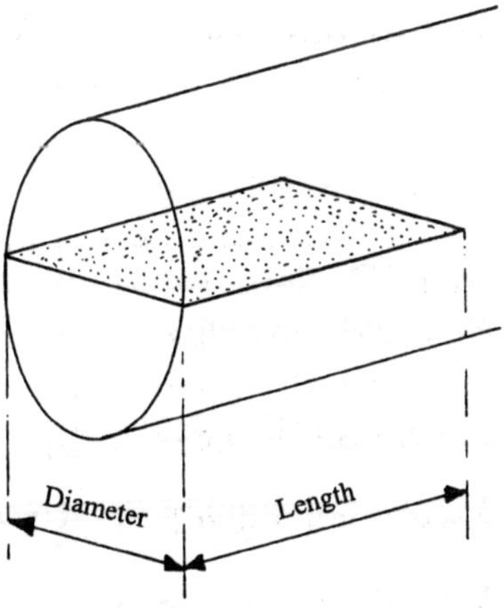

Bearing Area

diameter times the length of the supported section.

Diesel engines may use bronze bushing inserts and higher pressures. Pin diameter may be determined by the maximum allowable ovalization during firing and should not exceed .001 inch. Ovalization of the pin is determined by: *Howarth

$$\mathbf{0.04\mathit{l}\,(D^2pd^3)\,/\,E\mathit{l}t^3}$$

D= bore in inches, **p** = maximum cylinder pressure, **d** = pin diameter in inches, **l** = length of pin, **t** = pin wall thickness, **E** = Young's modulus (steel, 30,000,000 psi)

The center of the piston pin may be located .02 to .04D above the center of the piston to offset the turning effect of friction. In order to reduce piston slap, pins may be located slightly to one side of the piston axis. The idea being that the piston will rock when the pressure on the head is low and not when the piston is under high pressure at top dead center. The usual offset is 1.5% of the bore in the direction opposite the engine rotation.

PISTON SKIRT: The length of the skirt below the ring section should be such that the side thrust from the connecting rod does not exceed 25 psi during the expansion stroke. The **side thrust is determined by** :

$$F_{\text{side thrust}} = (F_{gas} + F_{inertia}) \times \{\sin\theta / \sqrt{(L/R)^2 - \sin^2\theta}\}$$

F_{gas} = the gas pressure and may be estimated from an indicator diagram.

$F_{inertia}$ = inertia force

$$F_{inertia} = -0.0000284\,W_i\,RN^2 f_a$$

W_i = reciprocating weight = piston assembly and top of the rod

f_a Crank Angle Factors for Piston Acceleration

Values of L / R

Crank Angle, Degrees		3	3.2	3.4	3.6	3.8	4	4.2	4.4
0	360	1.333	1.323	1.294	1.278	1.263	1.250	1.238	1.227
10	350	1.298	1.279	1.261	1.246	1.232	1.220	1.209	1.198
20	340	1.195	1.179	1.165	1.153	1.142	1.131	1.122	1.114
30	330	1.033	1.022	1.013	1.005	0.998	0.991	0.985	0.980
40	320	0.824	0.820	0.817	0.814	0.812	0.809	0.807	0.806
50	310	0.585	0.589	0.592	0.595	0.597	0.600	0.602	0.604
60	300	0.333	0.344	0.353	0.361	0.368	0.375	0.380	0.386
70	290	0.087	0.103	0.117	0.129	0.140	0.151	0.160	0.168
80	280	-0.139	-0.120	-0.103	-0.087	-0.073	-0.061	-0.050	-0.040
90	270	-0.333	-0.313	-0.294	-0.278	-0.263	-0.250	-0.238	-0.227
100	260	-0.486	-0.467	-0.450	-0.435	-0.421	-0.409	-0.397	-0.387
110	250	-0.597	-0.581	-0.567	-0.555	-0.544	-0.534	-0.524	-0.516
120	240	-0.667	-0.656	-0.647	-0.639	-0.632	-0.625	-0.619	-0.614
130	230	-0.701	-0.697	-0.694	-0.691	-0.688	-0.686	-0.684	-0.682
140	220	-0.708	-0.712	-0.715	-0.718	-0.720	-0.723	-0.725	-0.727
150	210	-0.669	-0.710	-0.719	-0.727	-0.734	-0.749	-0.747	-0.753
160	200	-0.684	-0.700	-0.714	-0.727	-0.738	-0.741	-0.747	-0.765
170	190	-0.672	-0.691	-0.708	-0.724	-0.738	-0.750	-0.761	-0.771
180	180	-0.666	-0.688	-0.706	-0.722	-0.737	-0.750	-0.762	-0.773

Table 3

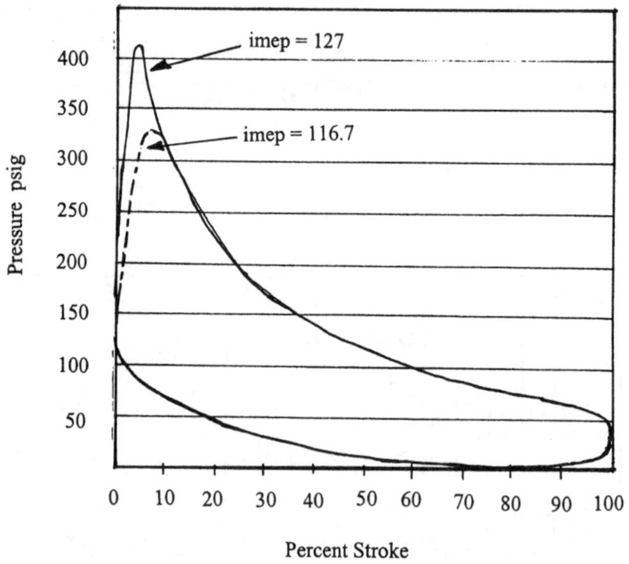

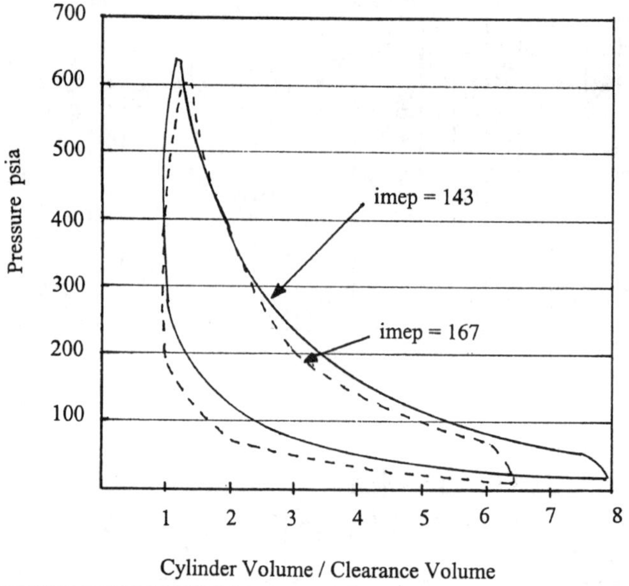

Typical indicator diagrams for 4 different spark engines. Naturally aspirated diesel engines will have peak pressures between 950 and 1300 psi.

To find the weight of either end of the rod, support the rod on knife edges at the centerline of the bearings with the rod being horizontal. The knife-edge at the end to be weighed rests on a scale. Verify the results by comparing the weight of the rod to the sum of the end weights.

N = rpm

f_a = the crank angle factor to piston acceleration. It is tabulated in table 3 for several common rod lengths or may be approximated by:

$$f_a \approx \cos\theta + (R/L)\cos 2\theta$$

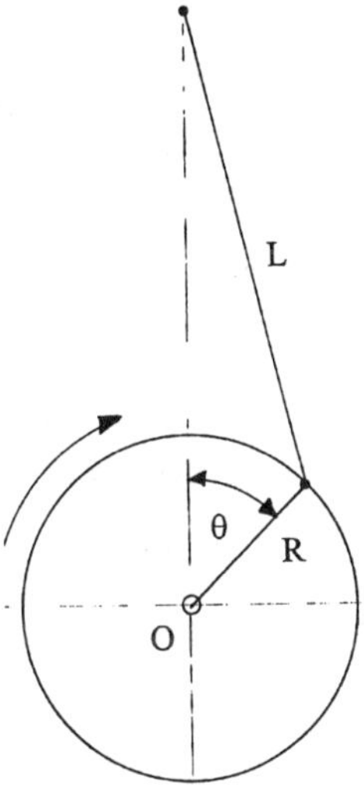

SURFACE FINISH: In order to prevent piston seizure, a film of oil must be maintained between the piston and the cylinder wall. Piston surface roughness values from 60 to 120 µinch (.00006 to .00012-inch) are preferred.

SKIRT OVALIZATION OR CAM:

Because the pin bosses deflect outward under high gas pressure, they are relieved giving the piston an oval shape. Many pistons are cast with recessed pin bosses. Others may be cut or ground to an oval shape. Generally .002 to .003 inch per side is sufficient. Short pistons may have the skirt at the pin boss cut away. Longer full skirt pistons have zero ovality at the base. The pistons are round at the bottom and the ovality increases up to the pin boss.

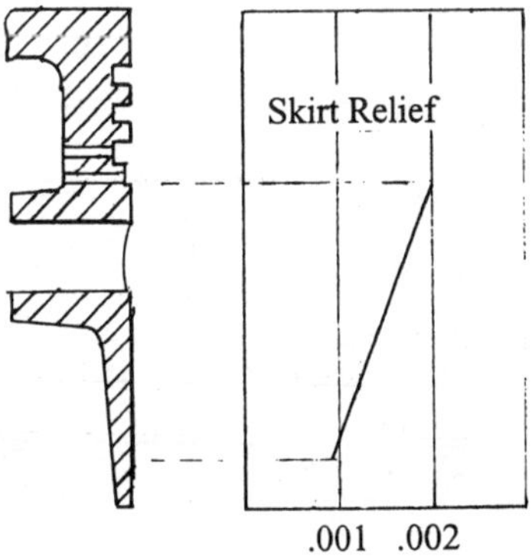

PISTON RINGS:

Piston rings are available in sizes from 2-inches diameter to 9 ¼-inches diameter. Smaller sizes may be available as service parts from various "weed-eater" manufacturers. Given the wide variety of standard and oversize rings, you should be able to bore your engine to fit one of the commercially available ring sets.

A few ring parameters are discussed below for those who may consider making their own. Piston rings are generally made of cast iron. Commercial casting of piston rings is described in *Metal Casting 2*. Model builders often choose to make their own rings by cutting them from cast iron stock, spreading them to form a gap and annealing. When cool, the rings have a permanent gap and must be compressed to fit into the cylinder. If properly made, the ring will then press outward against the cylinder wall with equal pressure around its circumference. The size of the gap is critical if the pressure is to be uniform. The gap and other ring dimensions are determined mathematically between the limits of the strengths of materials, and the required cylinder wall pressure.

If a ring is not to collapse under a vacuum, it must exert sufficient pressure against the cylinder walls. Engines routinely pull 8.5 to 10 pounds of vacuum between closed throttle idling and high-speed operation. Using a safety factor of 3, the wall pressure should be at least 30 psi. But not so high that the ring is liable to break in operation. These factors dictate both the radial thickness of the ring and the gap size. Wall pressure is governed by the radial thickness of the ring. Thicker rings create higher wall pressure but are also more difficult to install. The optimal thickness is .045 times the bore diameter. SAE specifications spread it between .041 and .046 as required for ease of installation among other things. The optimal gap is .155 times the bore diameter. Close inspection of many commercial ring sets reveals that they closely conform to the given ratios.

Ring height varies between .0189 bore to .046 bore. Rings of large diesels may be as short as .17 bore. Frictional losses are smaller when using the shorter rings. Automotive applications tend to use the smaller heights. Thicker rings are more abrasion resistant and used in applications such as small industrial motors and chainsaws. Smaller diameter rings may also be easier to handle when they are higher.

Installed ring gap: To prevent seizing of the ring due to expansion of the ring at operating temperature, an additional end gap must be provided. SAE recommends .004-inch gap per inch of cylinder diameter. Others specify .005-inch gap per inch of cylinder diameter.

Summary of ring properties:

Thickness / Bore	.043	.045
Gap / Bore	.155	.155
Wall Pressure	40psi	46psi
Operating Stress	59700psi	62800psi
Installation Stress	73400psi	83500psi
Installed Gap / Bore	.004-.005	.004-.005

Expanding the ring and shape sensitivity: For the ring to compress back to a perfectly round shape, the spreading force must be applied perpendicular to the gap faces by a tapered wedge or a dowel. The open rings are clamped flat in a fixture to prevent warping and are then annealed.

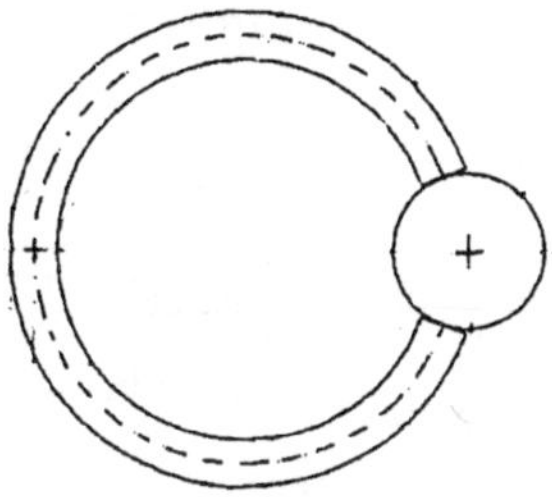

The spreading force should be applied at the centerline of the ring thickness.

Commercial Piston Rings:

Seen below is a partial listing of available single cylinder ring sets. Sizes over 4 ¼-inches diameter are usually multiple cylinder sets. There are many sizes, both metric and standard between the sizes listed below. The following part numbers are for piston rings (3 ring sets) that are 3/32, 3/32, and 3/16-inches high. Additional rings may be found which are 1/16, 1/8, 5/32, and ¼ inches thick. You should consult the factory for additional part numbers. They can usually make a custom set if you are unable to find what you want as a stock item. Rings as large as 9 ¼-inches diameter are available.

Hastings Piston Ring Part Numbers			
Bore	Part Number	Bore	Part Number
2	240, 2C7894	3 3/16	4663
2 3/32	406	3 1/4	6535
2 1/4	413	3 3/10	4230
2 5/16	7798, 2C7798	3 5/16	6627
2 3/8	236, 2C7576	3 3/8	2C7144*
2 1/2	7898	3 7/16	6588
2 9/16	6473	3 1/2	6427
2 5/8	7889, 2C7889	3 9/16	6962
2 3/4	6008*	3 5/8	300
2 13/16	238	3 11/16	2C6015
2 7/8	295	3 3/4	2C5466
2 15/16	4796	3 7/8	6353*
3	2C7502	4	2C6314
3 1/16	4474	4 1/4	2C6531
3 1/8	235		
Rings are 3/32, 3/32, 3/16 in, 2C denotes chrome ring			
*6008 -2 cyl set, 2C7144-4 cyl set, 6353- 6cyl set			

Hastings Manufacturing Co. (269) 945-2491

Filters and Runner Traps:

Filters are very effective in removing dross from aluminum castings. They are best when placed close to the casting, however they work well when placed at the bottom of the sprue. I have never had a piston casting rejected because of dross inclusions when using filters.

Sheet filters are available in several mesh sizes, and they are relatively inexpensive and easy to use. They do not require prints and may be inserted between the cope and drag at the gates or the sprue base. Sheet filters are easily trimmed to size using scissors. Currently the cost is approximately $8.00 for a 12 x 12-inch sheet from which many filters may be cut. The 2-inch square filters seen below are supplied by Ametek.

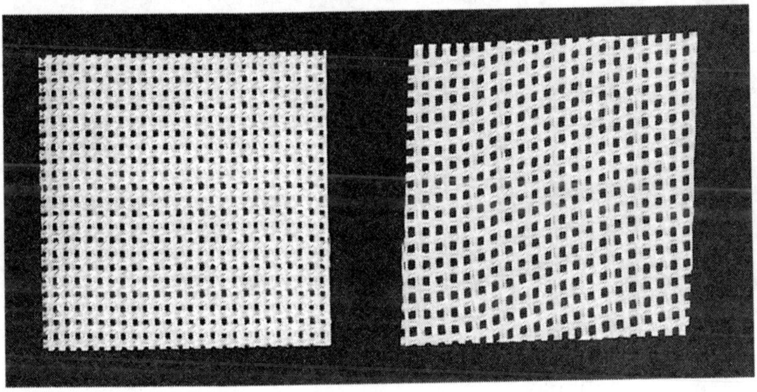

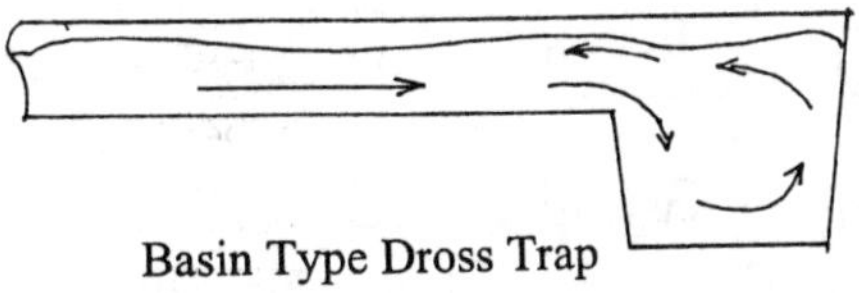

Basin Type Dross Trap

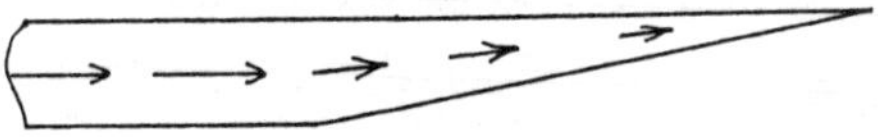

Tapered Dross Trap

Runner extensions are used after the last ingate to trap the first metal into the system because it usually has an accumulation of dirt, gas and dross. A few inches are usually enough and the end of the runner is vented so that gas pressure will not prevent the extension from properly filling. Tapered dross traps are preferred because the metal freezes in the tapered section preventing contaminants from washing back into the runner. Basin type dross traps cause a circulating flow.

Piston blank casting with tapered dross traps at the end of the runners and a sheet filter at the base of the sprue.

Filters are not essential for casting pistons, however they will certainly reduce your scrap rate. Filters also improve both the pressure tightness and mechanical properties of the casting by reducing the number of entrained oxide films (dross).

POURING, FEEDING AND SOLIDIFICATION OF PISTON CASTINGS:

Pistons may be made of cast iron or aluminum. Iron pistons are easily cast by using small gates on the top edge of the casting and no risers. Aluminum pistons, because of solidification shrinkage, are more difficult to cast. Pouring temperature and the placement of gates and risers are very important.

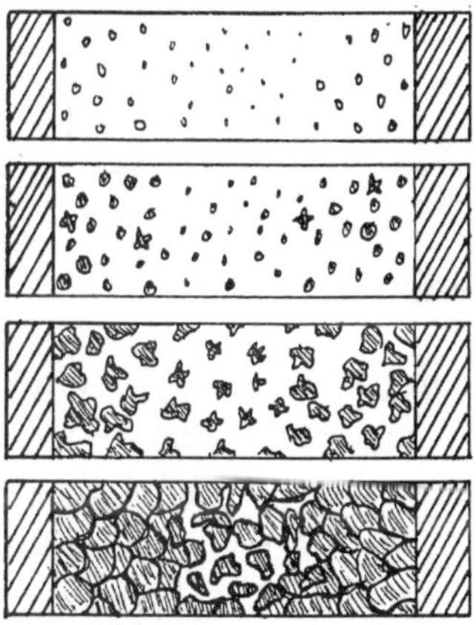

Long Freezing Range Alloy

Aluminum castings freeze by three different methods. In pure aluminum, shrinkage occurs as a deep pipe or at the centerline of the casting. Solidification of alloy #295, 94% aluminum, 5% copper, 1% silicon begins at the wall but progresses quickly to the center of the casting. Fine grains form randomly in the center of the casting and freezing continues in a mushy state. The center of the casting may be as much as 85% solid before a completely solid skin forms on the surface. As a network of solid grains form, feed metal is unable to flow through the constricted passages and microshrinkage occurs around the dendrites. The riser height drops and distributed microshrinkage forms throughout the riser and casting.

Chills are used to force the metal to freeze quickly from one end before the network of grains forms, constricting the flow of feed metal. Chills also increase the mechanical properties by reducing the segregation of gas and impurities at the grain boundaries.

Many pistons are cast from alloy F 132, or #332 (they are equivalent alloys). Alloy #332, silicon 9.5%, copper 3%

solidifies with some gross shrinkage and some distributed microshrinkage.

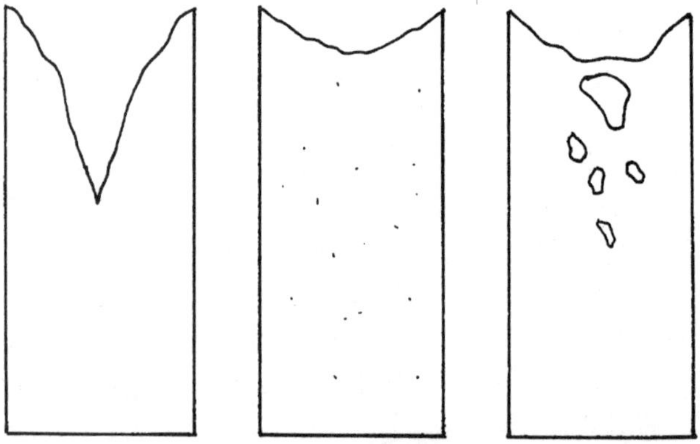

Deep Pipe Dispersed Shrinkage Combined

RISERS AND FEEDING OF CASTINGS:

Because long and short freezing range alloys solidify differently, no one set of specific guidelines can be given for the placement of all risers. General riser dimensions are given but should be modified to suit the particular job at hand. For the small foundryman, selection of proper risers is still a trial and error affair.

Guidelines that generally represent the short freezing range or skin forming alloys have been generated by years of experience in steel casting. In these alloys, shrinkage occurs as riser piping, gross shrinkage at hot spots and centerline shrinkage in uniform sections. For this situation, use hot risers gated directly from the runner when possible

Many aluminum alloys are not skin forming but freeze in a mushy or pasty state with dispersed micro-shrinkage. These alloys behave differently than short freezing range alloys. *Heavy risering may not significantly improve the situation and may make it worse.* Good feeding is better produced by steep temperature gradients towards the riser. This is accomplished by proper placement of chills and insulating boards. In some situations, micro-porosity is not a problem and the foundry seeks to distribute the porosity as widely as possible throughout the casting. This is accomplished by making it solidify as uniformly as possible. When section thickness is mixed, gating into thin sections, the placement of chills and dead risers on the heavy sections helps reduce the sink marks or depressions. The dead risers still must remain liquid longer than the casting so they should be insulated or topped with hot metal.

Piston castings have two areas where the increased section thickness may cause hot spots and internal shrinkage, at the pin bosses and at the point where the gates join the casting. Heavy risering does not appear to help the situation. Generally, I prefer using chills to encourage directional solidification however; this complicates the molding for a short run part. Gating and risering at the pin boss has never produced a sound casting due the large increased section thickness. I have obtained the best results by gating into the thin sections located 90° from the pin bosses. *The highest number of good castings results from using a combination of small gates and low pouring temperature.*

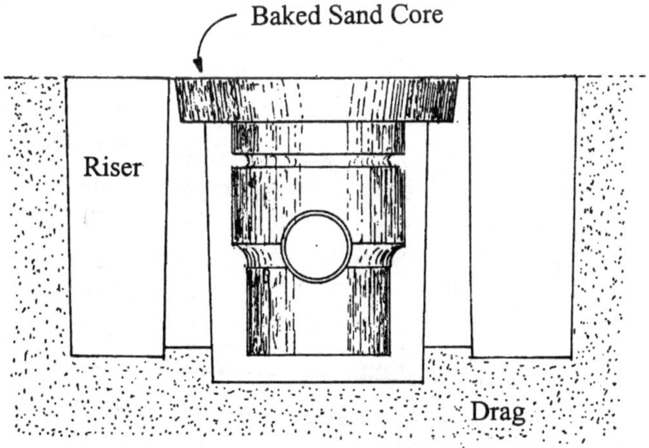

Gating Scheme 1. The gates must be no thicker than .6 the wall thickness

In order to prevent hot spots from forming where the gates join the casting, the gates must be thin, similar to those used for plate castings. The maximum gate thickness is approximately .6 the plate thickness. The risers shown on the drawings are not intended to feed the casting, but to feed the gates so that they do not draw metal from the casting wall. Long, thin, tapered gates are somewhat difficult to make. A second and simpler scheme is to use very short risers and gate into the thin top section of the (inverted) casting. I recommend starting with this scheme.

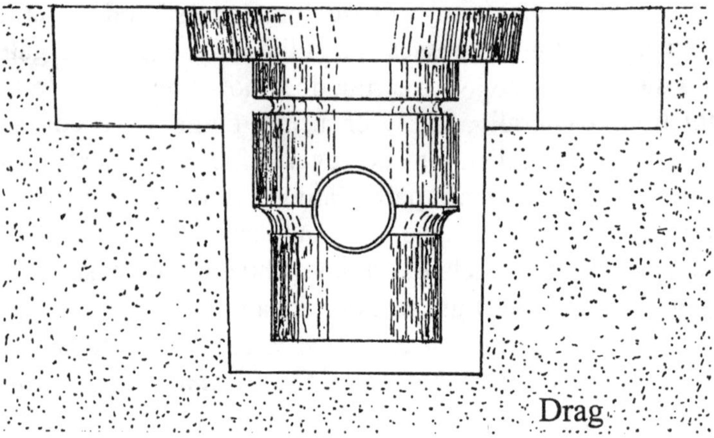

Modern pistons are most likely cast from alloy #332 or #336, both of which are permanent mold alloys. They have a high silicon content making them very fluid. The solidification range of #332 is from 1080 to 970°F, and the solidification range of #336 is from 1050 to 1000°F. When using these alloys, the best sand cast pistons are made when the pouring temperature is approximately 100 to 120° F above the solidification temperature. Gross shrinkage is seen in piston 1 (next page). It was poured at 1350°F. Piston 2 is the same mold poured at 1200°F.

Scrap pistons may be melted for casting alloy if you first pour ingots. This removes the dirt, oil and water from the alloy that causes gas defects.

Piston 1- poured at 1350°F

Piston 2- poured at 1200°F

MAKING REPLACEMENT PISTONS:

SEQUENCE OF OPERATIONS:

1. Measure the Bore
2. Find Rings
3. Make Drawings
4. Make Patterns
5. Cast Piston Blanks
6. Bore Reference Surfaces
7. Make Piston Mandrel
8. Perform Lathe Operations
9. Perform Mill Operations
10. Create the "Egg Shape"
11. Remove Turning Boss and Balance

1. MEASURE THE BORE

In order to select piston rings, you must inspect the cylinders for damage and determine if and how much material must be removed to clean up the walls and present a proper surface. Most cylinders will clean up when bored .020 to .030-inch oversize (increase in diameter). Standard oversize ring diameters are .020 and .030-inch; however you may often purchase oversize rings at .040, .060, .080, .100 and .120.

You may find that your engine has already been bored oversize. If you are unable to find a standard oversized ring set to match your new cylinder diameter, you may be able to go to the next larger sized standard or metric bore. For instance, your standard bore is 3 7/16-inches and it has been bored .040 over. You are unable to find .060 rings; however you may be able to go to 3 ½-inches and use a standard set of rings. Older engines usually have thick cylinder walls that may be bored well over their original size. If you have a bad cylinder that will not clean up without excessive boring, you can always insert a cylinder sleeve.

Cylinders are best finished using a "Sunnen Type" cylinder hone or equivalent. Leave this job to an automotive machine shop. Properly finished cylinders will have a cross-hatch pattern at a 44° to 62° angle. The cross-hatch surface holds oil required for proper lubrication and sealing. If your engine does not have hardened valve seats, you may cut the top of the block flat using a face mill. Otherwise, you should have the top of the block surfaced when you take your engine to the machine shop for honeing. Be sure that the top edge of each cylinder is chamfered, or you will have trouble installing your pistons.

2. FIND RINGS:

You do not have to use the same size or type of rings on your new pistons. Your pistons may have cast iron rings and a cast oil ring. A modern set may have chrome rings and thin spring steel oil rings separated by a spacer. The rings do not have to be the same thickness. Unless you are building a diesel, you are really only concerned with getting the proper bore size. You are making the pistons and you can make them any way you want! I built one ring groove cutter that I use on all of my pistons. All of the pistons seen on the front cover have the same thickness of the ring grooves. Spacing is easily changed using shims. Although it would be nice to have a metric groove cutter, I take my SI engines to the closest "inch" size and cut the same grooves.

Hasting's part numbers for several inch-type single cylinder ring sets are listed in the appendix. The thicknesses of the rings are 3/32, 3/32, and 3/16-inch. There are other sizes available. You should consult your auto parts supplier, request a ring catalog or call technical support at one of the ring manufacturers. Currently, I am paying about $12 to $15 per cylinder for rings.

3. MAKE DRAWINGS:

Using a dial caliper to accurately measure a cleaned piston, make a full sized drawing of your existing part. You must accurately record and sketch all of the information. Because you will constantly refer back to your drawing during both the pattern making and machining processes, you might want to make a few photo copies. I find myself writing notes all over them as I calculate machining distances and thicknesses.

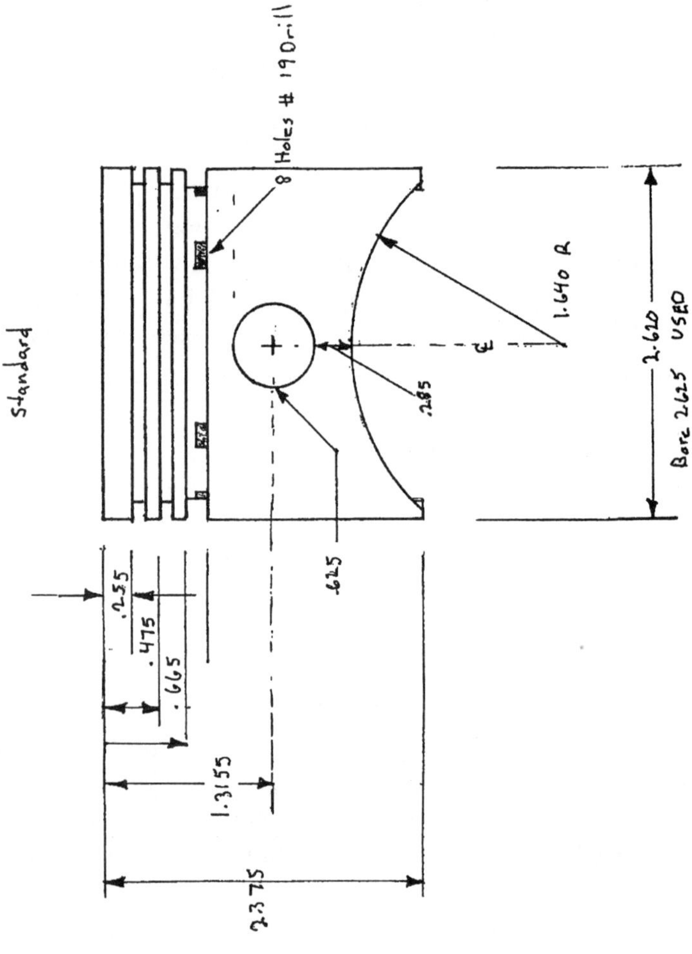

A typical piston drawing seen on previous page. By making a good scale drawing you will observe the difference in diameter between the pin side of the piston and the thrust side. You will also notice the extra clearance at the top of the piston required for expansion as the piston reaches operating temperature.

You should probably make the complete drawing of your first piston. After that, you may choose to skip the detail drawing and move directly to the piston blank drawing as seen on the next page. Note that if you make the ring groove cutter, you can skip the ring spacing information in your drawings because it is predetermined by the cutter or the radial thickness of the rings (page 13).

After the part drawing is made, make drawings of the core and piston blank pattern. This can be a tedious process. You must add machining allowances and draft to all the sides while maintaining the proper thickness of the head and ring belt. Because sand-cast pistons will have lower mechanical properties than permanent mold cast pistons, remember to add .050 to .1-inch to the wall thickness (smaller inside diameter of the piston). Placement of the pin bosses is also important. Finally, **add a rib that runs around the inside bottom of the piston**. When machining, all of the lengthwise dimensions are located relative to the bottom surface of this rib. If you locate it the same distance relative to the pin bosses on different core molds, you can use the same mandrel for turning several different types of pistons.

Machining allowance: The piston is finished relative to the core, and there are several situations where the core may become slightly misaligned in the casting. The rough casting must also be chucked in the lathe and a reference edge is cut relative to the core. Because there are many opportunities for error in these two processes, you should add approximately .175 to .25-inch to the outer wall thickness. This will increase the diameter by .35 to .5-inch.

After you have made a few pistons, you may find that you can use less machining allowance.

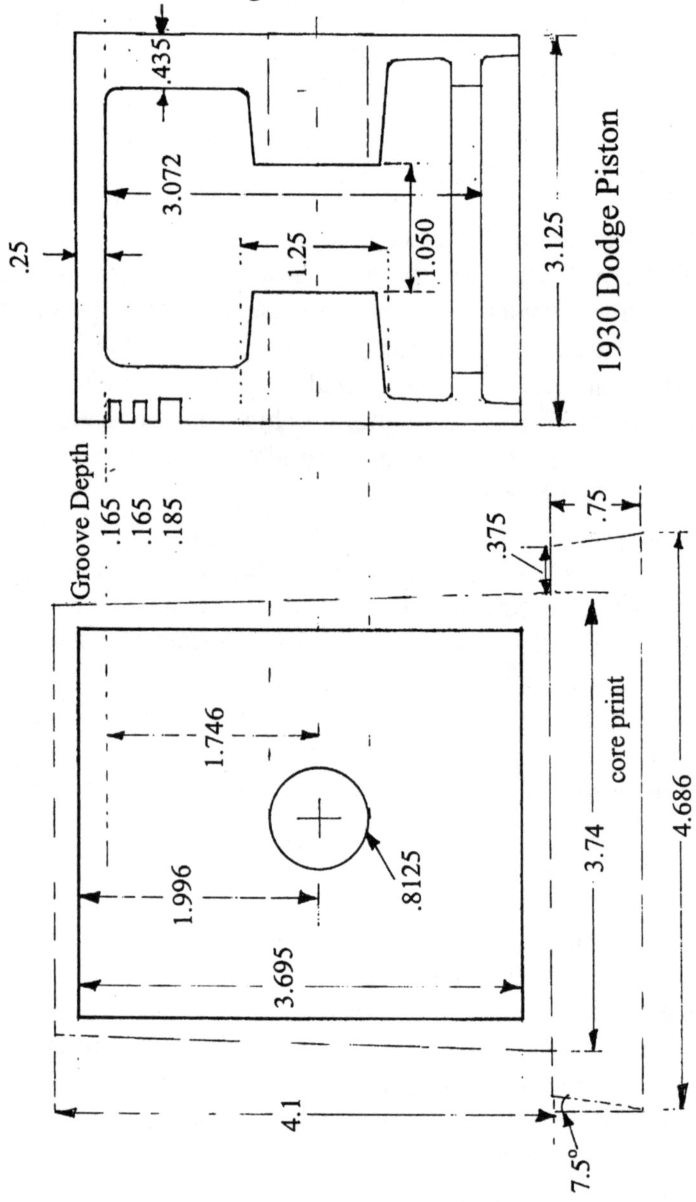

Piston and Pattern Drawing

Using too little machining allowance saves neither time nor metal, as a higher percentage of castings may not properly clean up. The machining allowance is seen on the drawing of the preceding page. The core print seen on the bottom of the piston, is turned from a .75 inch thick section of wood. It has a 7½ $^{\circ}$ taper. The piston body has a 1½$^{\circ}$ degree taper. The pin bosses in the core-box have a 5° taper.

4. MAKING THE PATTERNS:

Pattern Wood: Mahogany machines very well and is the best pattern wood, however it is expensive. Yellow pine is very inexpensive, readily available and will make workable patterns, but has a few drawbacks. It should be dry before working or it will change dimension quickly, frustrating any attempt at precision (another good reason for using a large machining allowance). Purchase yellow pine several weeks before you start your project and allow it to dry in your shop. The grain of yellow will rise upon shellacking, requiring much sanding to get a smooth surface. While the cylindrical blank patterns are easily smoothed in the lathe, the inside surfaces of the core box must be smoothed by hand, which is a time consuming process.

The core-box is the most difficult part of the pattern project, so make it first. If the finished dimensions of your core box are a little off from your drawings, it is easy to make adjustments to the cylindrical piston blank. You can make the blank pattern match the core-box easier than you can make the core-box match the blank pattern.

The core box is split down the middle with one pin boss in each side. Larger cores might be made in halves and glued together. Very large cores may be made as rings and bolted together. This type of piston core is used later in the *Small Foundry Series* when a 10-inch diameter jolt-squeezer piston is cast.

Split Core Box Made From Yellow Pine

Plane and glue up sections of wood until each edge is at least ½-inch larger than the core print cutout. See the photo above. When the glue has dried, plane the sides flat and saw the board in half (half as long to form two short sides). Clamp the halves together and cut the ends so that the blocks are exactly the same length and square.

Making the Dowel Holes: Select two close fitting edges for the parting line. Drive two brads or small nails into one of the parting-surfaces. Be sure that the brads are perpendicular to the surface and not bent at an angle. Using a hand grinder or sturdy snips, cut the brads off and grind or file them until they protrude 0.1-inch from the surface. Round the corners of the nails. Carefully set the mating surface against the nails, being careful to keep the blocks square. Squeeze the blocks together in a vise or rap the back of the wood block with a hammer to mark the location of the holes on the mating piece of wood. Using a number D drill (.246-inch), drill ½-inch deep holes at the center of the marked locations to mount the alignment pins (dowels). Cut a 5/8-inch length of ¼-inch diameter brass rod (or dowel) and round the edges. Put glue into the hole and drive the dowel down until approximately 0.175-inch protrudes from the surface.

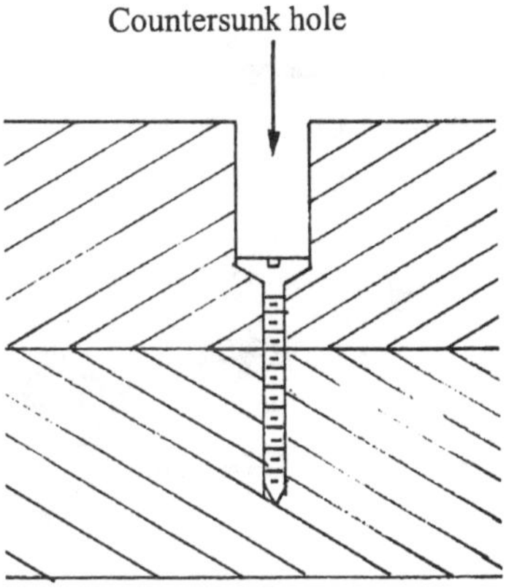

Remove the brads from the remaining block and drill the holes with a #G drill. The holes need only be a little deeper than the protruding dowels.

Clamp the blocks together and drill holes to accept long drywall screws (notice the small holes in the photo of the core box). You will most likely have to counter sink the holes for the screw heads an inch or so deep. Insert drywall screws to hold the blocks together and transfer the assembly to the lathe.

Drilling the Center of the Core Box

Carefully center the assembled blocks in a four-jaw chuck. Face one end square, then drill and bore the smaller diameter of the piston core through the center point of the parting line. Using a shop-vac to catch the wood chips at the tool bit makes this dusty job much more pleasant.

Boring the Core Box

Bore the lower part of the piston diameter to length. Bore the core print to proper size and taper the sides at 7½°

At this point you have to make a decision. You can choose to finish the box as it is or you can bore it a little oversize, give it a few coats of Bondo auto body filler and bore it back to dimension. This leaves a very smooth surface that requires little finish sanding. Shellac raises the grain on bare yellow pine requiring time consuming finish sanding. The inside of the core box must be *very smooth* or the cores will slump and distort when you try to remove the core box. You must slice the parting line with a razor to get a clean break in the Bondo when opening the freshly coated and bored box.

* Although I rarely do, at this point, a smart fellow would cut a groove for the rib located at the base of the piston. Later, the rib can be cut from ¼-inch thick stock and inserted into this groove. (Hindsight is always 20-20.)

Remove the core box from the lathe, turn the box over, square and center it in the chuck. Face the box to length as required for your piston. Using Bondo, fillet the joint between the two different diameters.

Open the box and move it to the mill. Carefully measure down from the head of the core box to locate the center of the piston pin boss. Drill with an end mill because other types of drill bits tend to wander off center or produce an oversize hole. For smaller pistons, I use a 1-inch diameter end mill. For larger ones, I use 1¼-inch diameter or larger. These holes must be straight and square to the inside surface of the core box, or later you will have trouble drawing the box from the green sand core.

Locating and Drilling a Hole for one Pin Boss
Note that the rib has been inserted

Cut the pin bosses, with a 5° taper, in the lathe. Make the shanks a snug fit, not loose and wobbling in the pin boss holes. Coat them with glue and press them into position. Work quickly or the glue will set and lock the bosses in place. Pay particular attention to their height above the inside surface of the core box. If you are not careful, your pin bosses may be too long requiring you to trim the castings later. When the glue has dried, fillet the joint between the bosses and the inside surface of the box with Bondo.

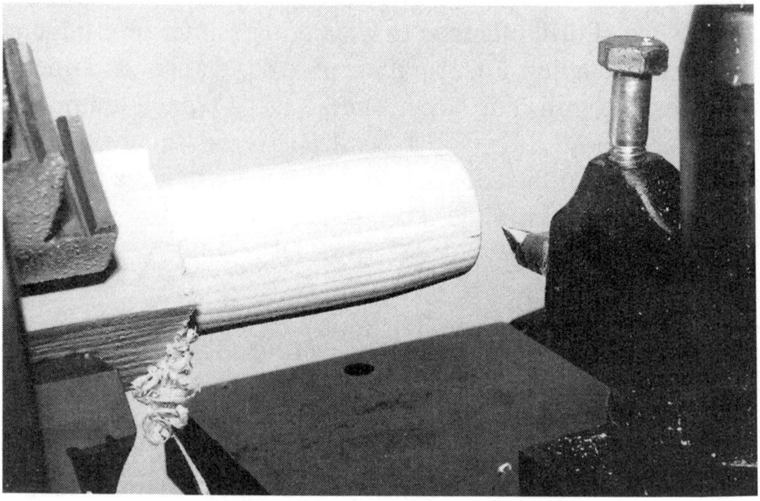

Turning the Taper on the Pin Bosses

Sand a slight taper on the top of the core box, approximately 1-1½°. Cut strips from ¼-inch plywood for the top. Glue and insert brads to hold them in place. If you are using any ribs between the piston head and the pin bosses, insert them now. The ribs may be cut from ¼-inch plywood. Fillet them and sand smooth being sure that there are no undercuts to cause the core to hang in the box and distort. Sand the inside surface of the core box. Paint the

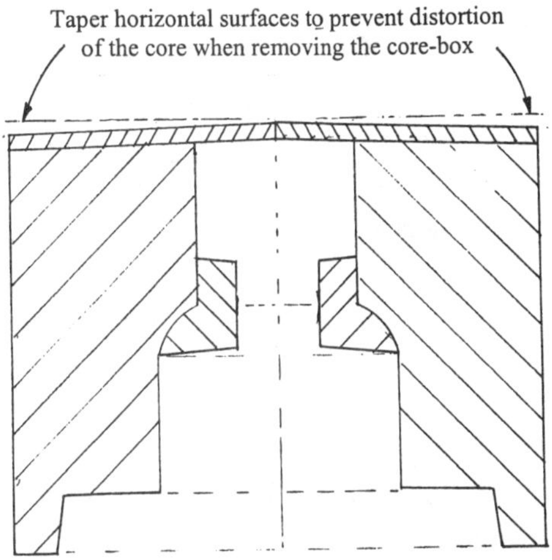

core box with two coats of shellac. Sand the inside smooth and rub it out with steel wool for a very smooth finish.

Making the Piston Blank Pattern: Turn the piston blank in the lathe. Mount the two sections of pattern wood on a mandrel as seen below.

Turning the Piston Pattern in the Lathe

The bottom section is a rough sawn disk, .75-inches thick. The top section is glued up from several strips of yellow pine. To reduce the time and mess of turning the blank

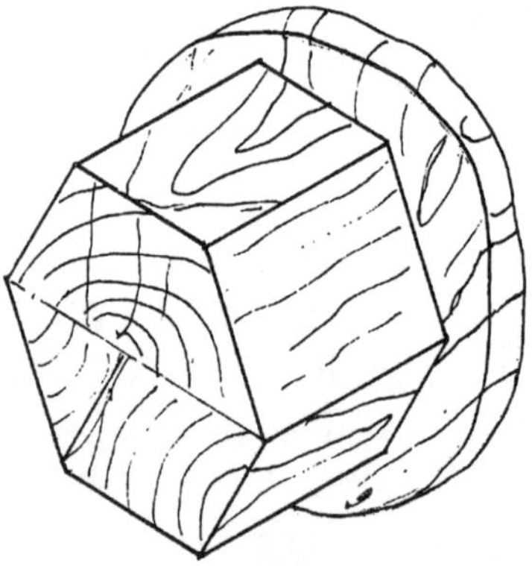

round, the corners are sawn off the block forming an octagon. The octagon is center drilled and glued and screwed to the bottom section that forms the core print. Turn the blank with a 1½° taper and the core print with a 7 ½° taper.

Make the mandrel by welding a section of plate to a short section 1-inch diameter steel rod. Turn it round and face the plate flat in the lathe. Center drill the face of the mandrel, followed by a ¼-inch diameter hole approximately ½-inch deep. Drill 3 or 4 holes around the plate for mounting screws. Turn a sharp point on a ¼-inch steel rod and trim the rod so that the point will protrude from the hole in the mandrel. This point pierces the center point of the pattern.

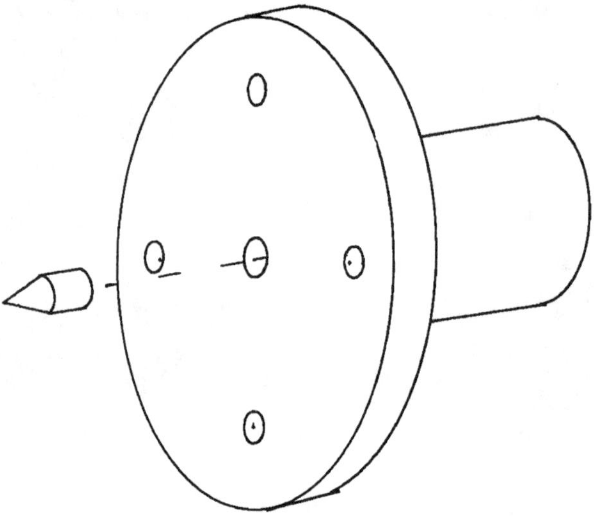

Piston Blank Mandrel

Mounted or loose patterns: Depending upon your flask and equipment, the patterns may be loose or mounted. Mounted patterns with formed gating systems produce cleaner and more accurate molds and castings. Mounting, however, is not essential, as I have produced many castings from loose patterns and hand cut gates. If you are going to produce more than a very few castings, mounting is best.

Mounted patterns require flasks that have straight pins located on the drag. If you are following along in the *Small Foundry Series,* the flasks and pattern vibrator built in *Metal Casting 1* work well for piston castings. Because the drag must be deep, I have added an upset to the bottom of my wooden flask. The upset may be of wood or metal. It may be bolted to the flask or free. The upset seen below is cast from scrap aluminum and bolted to the flask. The corners are cut at 45° so that the flask can be opened. A small ¼-inch plywood shim is used in the lower left corner to get a tight fit at the opening on the right.

This home made snap flask works well with mounted patterns. Because it opens and is removed from each mold, you can make several sand molds from one flask.

A pattern vibrator must be used with plate mounted patterns. The pneumatic vibrator shakes the pattern, releasing it from the sand.

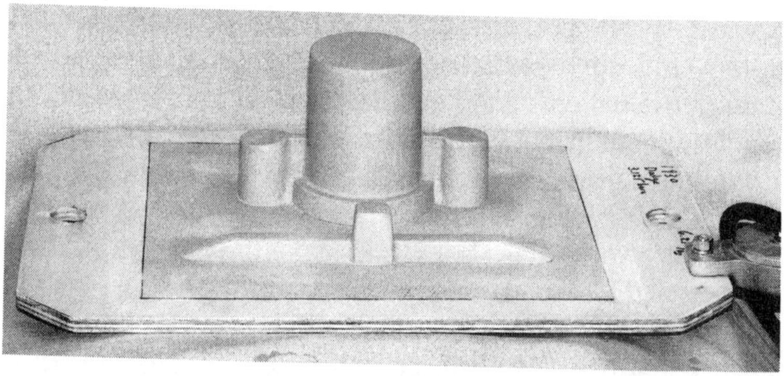

Plate mounted pattern with short risers and gating system. A pneumatic pattern vibrator is mounted on the right.

5. Casting Piston Blanks: Cast the piston blanks in green sand. Some may use Petro-bond sand, however oil bonded sands are very slow cooling and produce the lowest mechanical properties of all the casting methods. These castings may be heat-treated to improve both the mechanical properties and the machinability.

Core making: Make the cores from a mixture of sand, wheat or wallpaper paste, and molasses water. The sand fineness has a significant effect on the surface finish of the casting. Sands as fine as 150 produce a smooth finish, however they require more binder and do not vent as well. I often use a sand designated as "30-65" for my cores. It does not require as much binder, vents well and is very easy to remove after casting. It does produce a coarse surface finish and you must be careful that no loose grains fall off the core into the mold. Perhaps a #100 sand would be the best compromise between the two.

Clamp the finished core box together and fill it with dry sand. Pour the sand into a small bucket or large cup and using a small scale, such as a postal scale, weigh the sand.

You can now scale your mix up as required for several pistons with out shortage or waste.

Mixing 8 to 10 parts water to 1 part molasses makes molasses water. Using hot water and heating the molasses in a microwave oven makes mixing the solution easier. Any leftover solution may be stored in the refrigerator for several months if required. Add enough molasses water to the sand to make a slightly damp but not muddy sand and mix it well.

While you could use the wheat flour that you buy at the grocery store to mix your cores, much is required to develop a bond strong enough to prevent core slump. These cores are difficult to properly vent due to the high volumes of gas that are generated by the binder. Wallpaper paste is available at paint or Home Depot type stores and makes a very good binder. The paste may be the dry powder wheat or corn type. Little is required and either one works very well. Start with about 10% by volume, dusting it over the damp sand mixing thoroughly with a kneading motion. Do not dump it all in at once making a large blob of unmixed glue. Add molasses water and more paste as required to get a good bond. You want the core to hold together but not be doughy or form balls. Commercial machines mix the cores for 7 to 9 minutes so do not stop after 30 seconds and think that you have a good bond. Mix it well and add binder slowly. Give it a chance to absorb water and coat the sand grains. Initially, core mixing will be a trial and error affair. When you are satisfied with your results, write it down for future reference. Linseed oil may also be used as a core binder and is described in Metal Casting 1.

Coat the inside of the core box with paste wax or car wax. You may or may not rub it out. A rubbed out surface lasts longer but a wiped on surface is quick and easy. The wax prevents the cores from sticking in the mold. Dust the surface with parting dust and clamp the corebox together. Pour in a little core mix and using a dowel or pencil, ram it well into the corners. Add a little more mix and press it well against the sides above the pin bosses. Ram well against the outside surfaces to generate a good form. The center does not have to be rammed as tight so that the gasses vent freely. Poke a few vents into the core with a stiff wire or sharpened coat-hanger rod. Using a strip of sheet-metal, level the exposed surface of the rammed core. Place the sheet-metal over the open bottom of the core box and flip it over on a core drier (perforated flat metal sheet). Pull the sheet-metal out from under the core box. Tap the core box a few times to loosen the core and carefully open it. If your mix is good and the box is smooth with no undercuts, you should have a good-looking core. Make a few extra cores. You will need them to test the baking time and as spares if one breaks or distorts.

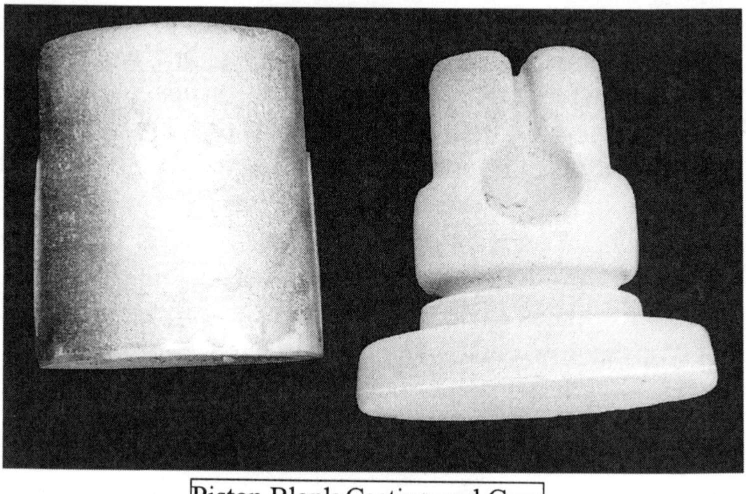

Piston Blank Casting and Core

Bake the cores at 325° F for 30 minutes. Remove a core and break it open to check the baking time. Continue baking as required to have a fully dried core. Record the time for future reference. I usually bake all of my cores for 30 to 40 minutes one day and set them aside. Later, I put them into the oven to finish baking while I ram the piston molds for casting.

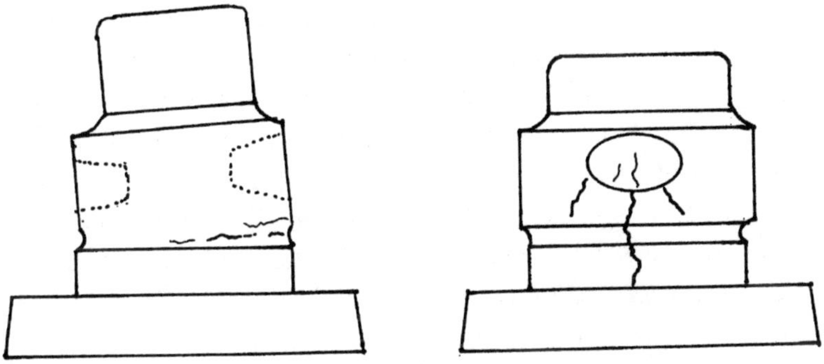

Typical Core Defects, Slump and Sag (exaggerated)

FINISHING AND INSERTING THE CORE:

After baking, check the cores for slump and squareness. Clean up any fins and rough edges with sandpaper. Blow off any loose sand. Because the core must be set in place with the pin bosses at 90° to the risers, grip the core base at the parting line (90° to the pin bosses).

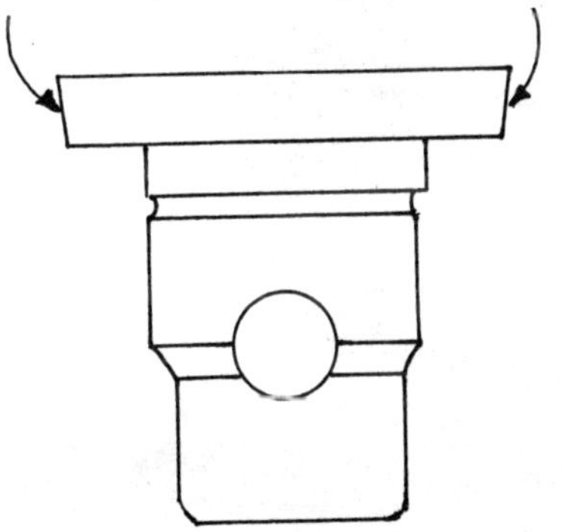

When inserting the core into the mold, the risers on either side of the mold should provide clearance for your fingertips. Melt and pour the castings as described in the Pouring and Feeding chapter. Pour at least 1 or 2 extra pistons. One piston can be used as a test piston for all the set ups and cuts, another may be required if you have a bad casting or machining error. No matter how careful I am, I always end up using a spare piston.

6. Bore the Reference Surface: All of the machining operations are located with respect to the lower rib that is cast into the piston. The piston skirt is bored up to the rib and a flat surface is cut on the bottom of the rib. The bottom surface is located a known distance from the inside of surface of the piston head. All lengthwise dimensions are calculated from this lower rib surface. The reference surface can be cut with a boring bar or by using the special cutter described below.

Left: Center the rough casting in a 4-jaw chuck using a surface gauge.

Above: Boring the bottom of the piston to create a reference edge

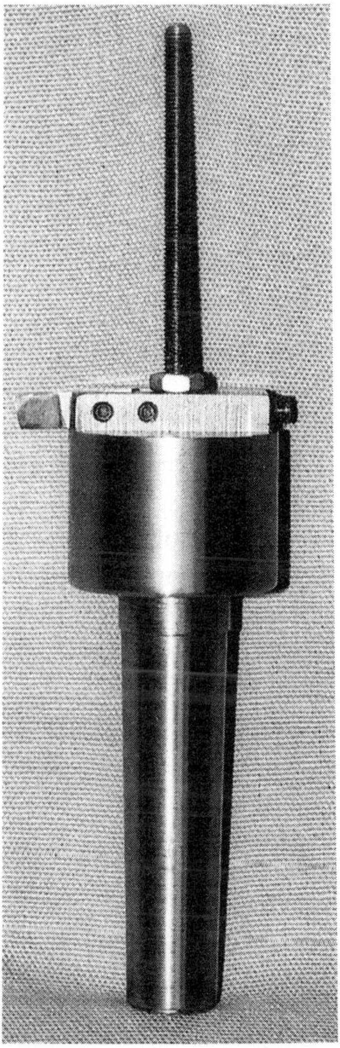

The piston skirt-boring tool seen on the **left** speeds up the machining of multiple pistons. The tool in the photo is good for pistons from 2 ½-inch diameter to 3 ¼-inch diameter. It is made from a scrap section of 2-inch diameter round stock. It has a #3Morse taper to fit the tailstock of my lathe. The taper is cut in the lathe by setting the compound rest to the proper angle. The angle is determined by holding a #3MT dead center between centers in the lathe and adjusting the cross slide using a dial indicator. The depth stop in the center of the tool is made from ¼-28 threaded rod. It works well for short lengths, however 5/16^{ths} or 3/8^{ths} rod would be more rigid when fully extended as seen in the photo. The tool bit is ground from 5/16^{ths} high-speed steel. The small setscrew on the right is the fine adjustment for the tool bit.

To use the tool, bore a piston using a boring bar as seen on the previous page. This allows you to determine the amount required to properly clean up the inside of the piston skirt. Insert the tool into the tailstock. Set the tool bit extension to make a light cut on the cleaned up surface. Next, measure the required depth from the piston head to the rib using a dial caliper. Set the depth of the stop so that

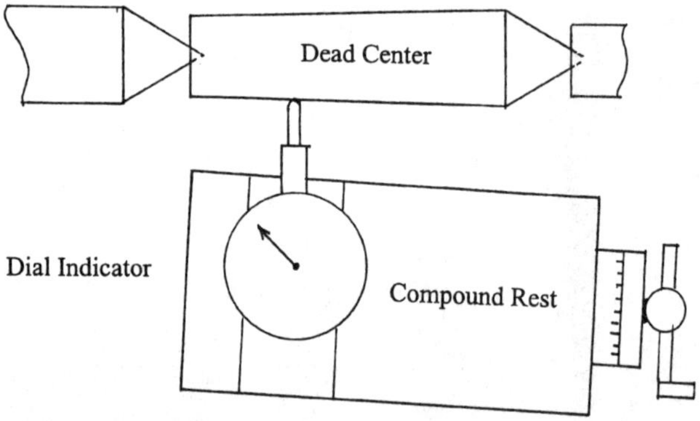

Setting the Compound Rest to Cut a Taper

a good flat surface is cut across the rib. Re-bore the piston using the tool and check that the diameter and depth are correct. Bore the remaining pistons by centering them in the chuck and running the tool up into the piston until it reaches the stop. This makes a quick job of boring multiple pistons. By keeping a boring bar attached to the tool post, you may quickly cut the inside diameter of each rib concentric to the cut made by the piston skirt boring tool.

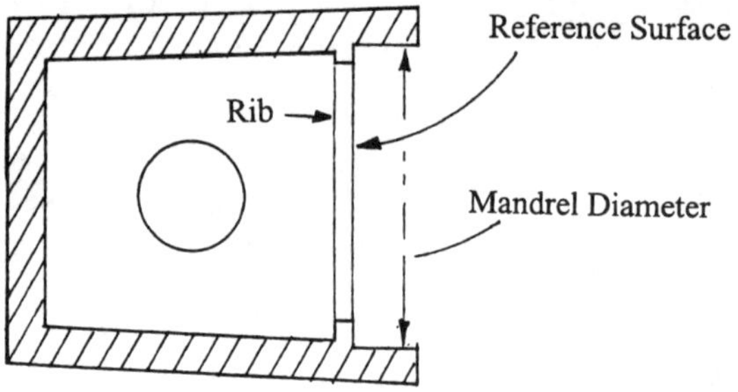

7. Make a Piston Mandrel:

The piston mandrel is a precision tool required to hold the pistons for machining in both the lathe and mill. It may be cut from steel, cast iron or aluminum for one-time use. Cast iron is preferred however most of my mandrels are steel.

Holding a section of round stock between centers, face the ends and turn the diameter to fit the inside of the piston

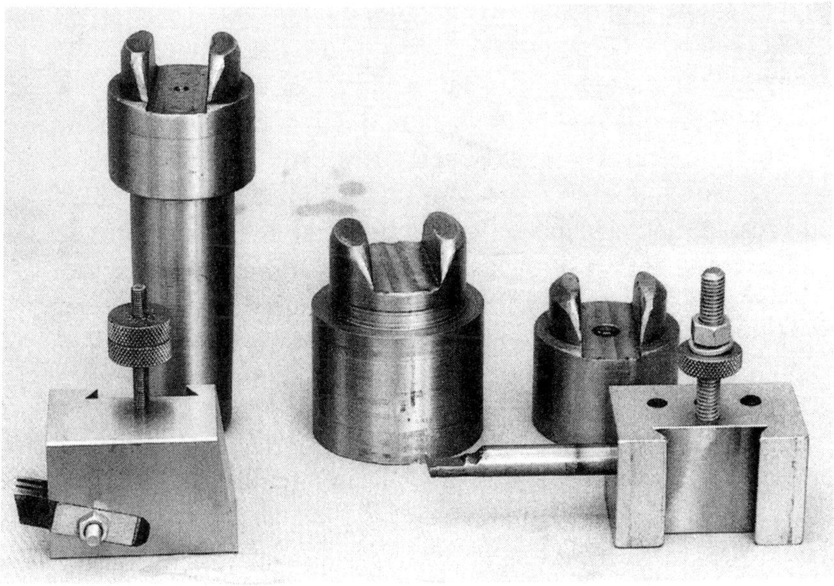

Left: Ring Groove Tool, Right: Piston-Pin Retaining-Ring Groove Cutting Tool, Rear, 3 different Piston Mandrels

skirt. For a 3.125-inch diameter piston, I am using a 3.625-inch length of stock. Turn a shoulder on the mandrel to accept the rib in the piston. Move the mandrel to the mill and cut a slot for the piston pin bosses. Using a hand grinder, round the edges of the mandrel until the piston properly fits down on the mandrel. This fit is critical for proper alignment and the process is a trial and error affair.

When you think that the piston fits, smear some bearing blue over the shoulder of the mandrel and press a piston down on it. The bluing should transfer all the way around the base of the rib.

If the bluing does not properly transfer, the piston does not fit squarely on the mandrel and the corners require further rounding. When things look good, center and square the mandrel in the lathe and recheck the fit. *All of your pistons will be junk if you neglect this fitting operation.*

8. Lathe Operations: The order of some of the operations is not critical but a matter of convenience. For instance, you may cut the piston to length before cutting the ring grooves, or you may cut the ring grooves first.

Check the lathe to be sure that it is not turning a taper. Press a piston blank on to the mandrel and center drill it. Secure it with a live center. Using a high-speed steel tool bit, turn the piston blanks approximately .080 oversize. This allows you to check for inclusions, scrap bad castings and prevents breaking a carbide tool bit on an interrupted cut.

Rough Turning the Piston Blank

If you are using a digital readout and a quick-change tool post, square the ring tool at the mandrel shoulder and zero the readout. Later, this allows you to quickly cut the ring grooves by changing tool bits. If you do not have these items, you will have to turn all the pistons and then lock the carriage in the proper position and cut one groove at a time on all of the pistons. This ensures that they are all in the same location on all the pistons.

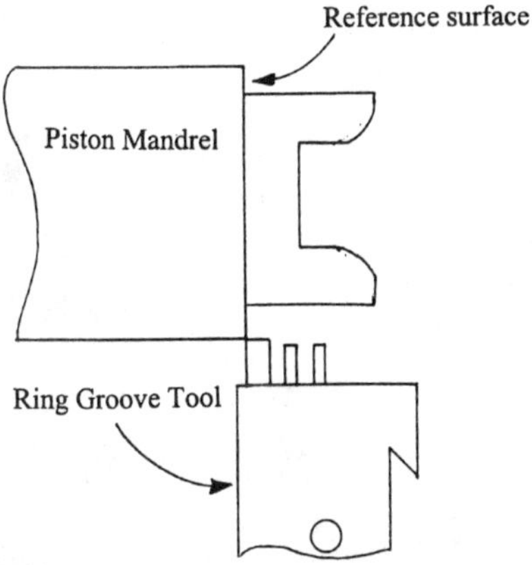

Modern pistons run with approximately .002 clearance on the thrust sides of the piston (as opposed to the pin sides). The pin bosses are relieved for an additional .004 to .005 clearance. Because the 1930's split skirt pistons are very flexible, they are fit with as little as .0006-inch clearance in order to minimize piston slap.

Clearance for All Aluminum Pistons					
Diameter (inches)	2 1/2 to 3	3 to 3 1/2	3 1/2 to 4	4 to 4 1/2	4 1/2 to 5
First Land	0.019	0.023	0.028	0.034	0.041
Second Land	0.014	0.016	0.019	0.023	0.028
Third Land	0.014	0.016	0.019	0.023	0.028
Fourth Land	0.014	0.016	0.019	0.023	0.028
Skirt	0.0035	0.0038	0.004	0.0045	0.0055

Clearance for Cast Iron Pistons					
Diameter (inches)	2 1/2 to 3	3 to 3 1/2	3 1/2 to 4	4 1/2 to 5	5 1/2 to 6
First Land	0.011	0.012	0.014	0.019	0.026
Second Land	0.008	0.009	0.010	0.014	0.019
Third Land	0.006	0.006	0.007	0.009	0.013
Fourth Land	0.006	0.006	0.007	0.009	0.013
Skirt	0.003	0.0035	0.0035	0.0048	0.007

Other "older" pistons were turned round using the clearance specified on the previous page.

Checking Final Diameter

Using kerosene as a lubricant, finish the pistons with a carbide or a high-speed steel tool bit having a 1/32-inch radius at the tip. Carefully check the final diameter of the piston after cutting approximately ¼-inch. Finish the piston diameter. At approximately 400 rpm, cut the ring grooves to depth using no lubricant. You may burnish the base of the grooves by applying oil and lightly pushing the tool bit back into the grooves. Trim the piston to length and face the head to the proper height. Remember all of these dimensions are measured relative to the shoulder of the mandrel. Leave a small boss on the head around the center hole.

Cut the proper relief on the ring lands as required for expansion at operating temperature. You may calculate the relief as described in the design section or use the chart

Cutting Three Ring Grooves

on page 53. Lightly round the edges and remove all burrs from the ring lands with a file while spinning the piston in the lathe at a low speed.

With all the ring grooves cut on one piston, install a set of rings and insert it into a cylinder. Any adjustment in ring depth must bc made before the oil holes are drilled because the ring tool hangs up and these holes making depth adjustment impossible.

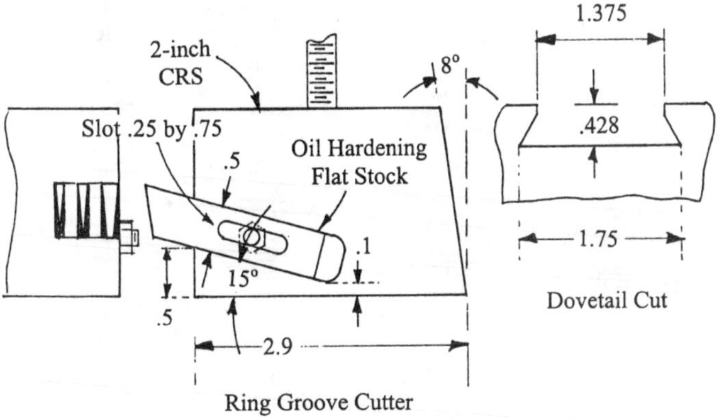

Ring Groove Cutter

You may cut the ring grooves with a single tool bit or with the ring groove cutter on the preceding page. The cutter is made to fit a standard dovetailed quick-change tool post. Shims between the cutters determine the ring land thickness. The 8° angle on the rear of the tool holder allows for quick sharpening on a surface grinder.

Piston Blanks with Lathe Operations Completed

9. Mill Operations: Prepare a pin-hole boring bar by clamping a section of aluminum in a vise and drilling a hole that is 1/32 to 1/64-inch smaller than the desired pin bore. Remove the drill bit and insert the boring bar into the spindle using a collet or a chuck. Set the boring bar tool bit to touch the inside diameter of the hole. Remove the boring bar and vise and bolt a dividing head to the table. Using a section of drill rod and a dial indicator, center and square the dividing head relative to the mill spindle.

Aligning the Dividing Head

When the head is true, center and square the piston mandrel in the dividing head chuck. Turn the mandrel so that the pin-boss cutout is vertical and centered with the spindle. Scribe a notch or use a sharpie pen to mark this location on the base of the mandrel. Turn the handle 20 revolutions and scribe another mark. This produces a visible reference to be sure that you are drilling the pinholes in the proper location.

Place a piston blank on the mandrel and secure it with the dividing head's tailstock. To prevent aluminum from building up on the bits, lubricate the part liberally with kerosene or WD-40 during all of the mill operations.

Hopefully all of your pistons are the same height from the reference surface, but in case they are not, *locate the pin hole bore relative to the piston head*. This maintains the compression height between all of the pistons.

Drill a ¼-inch diameter hole through one pin boss. Turn the crank 20 times, lock the dividing head spindle and drill the

other hole. Follow this with an end mill or drill bit to produce a hole for the boring bar. Bore the pinhole to size at approximately 600 rpm. Switch the mill to back gear (approximately 150 rpm). Lubricate the pinhole and ream the hole to size. You may use a reamer that is .001 under size or one that is the exact size of the pin. While under sized holes give a tighter running fit, I have had no trouble with piston pin holes reamed to exact size.

Reaming the Pin Boss

Usually, 10 holes in the oil ring groove are sufficient. However, you may add an additional relief under the oil ring with 8 holes. To prevent drilling a hole directly over the pin bosses, offset the piston 2 turns of the dividing head crank. Center drill the oil holes with 4 turns of the crank between each one (for ten holes). If you are using a 3/16^{ths}-inch oil ring, a #19 drill (.166) works well for the oil holes. If you are using an additional relief under the oil ring, turn the crank 2 additional turns to off set the holes. Follow the same procedure to drill 8 holes, skipping the hole directly over the pin boss on each side.

10. Ovalization or Creating the Egg Shape: To prevent the piston from sticking in the cylinder due to deflection at the pin bosses, they are relieved creating an oval shape. An additional .002 is removed from each side around the boss area. The relief tapers from a maximum at the pin bosses to zero at the bottom of the skirt of long pistons. The taper may be cut using a file or made with a flexible belt sander.

To cut the taper in the lathe, hold the mandrel in a 3-jaw chuck and secure the piston on the mandrel using the lathe tailstock. Put the lathe in a low gear so that the spindle will not turn while filing the piston. Without turning the lathe on, file the boundaries of the relief. Finish filing or sanding the inner section of the relief, checking the progress frequently with a micrometer. This is not a difficult operation and is not particularly a precision job. Inspection of several pistons reveals that many are completely cut away in this area.

11. Remove the Turning Boss and Balance:

Clamp the piston in a 3-jaw chuck and remove the turning boss on the piston head.

Usually, a set of pistons will come out within 3 to 5 grams of each other. Weigh each piston using a triple beam balance or equivalent and write the weight on the head of each piston with a sharpie marker. Select the lowest value as the target weight. Lightly clamp a piston in the 3-jaw chuck with the open end exposed. Cut away some of the lower rib. Weigh the piston and continue trimming until you reach the target weight. After a few cuts, you will quickly estimate the proper amount to remove making this a quick job.

Miscellaneous Operations: The skirts of some pistons are cut away to provide clearance for the crankshaft counter weights. This relief is easily cut in a mill as seen in the photo on the next page

Piston pins may be made from water hardening drill rod. Drill the center to lighten the pins then balance them using a triple beam balance. The hardening process may warp the pins; therefore I use the rod in the unhardened condition.

Ring clips at each end secure full floating pins. The clip retaining grooves may be cut in the mill using a boring head or in the lathe. Make a grooving tool from a section of drill rod. Harden by heating until non-magnetic and

quenching in water or oil as required. Clean the scale from the tool, temper to a light brown and quench. Grind as required for your clip.

Left: Cutting Crank Clearance

A piston is held in a lathe mounted piston vise to cut the pin retaining-ring groove as seen on page 4. The "V-block" vise is a shop made tool and **not required**. The jig pictured below is a quick and easy method of holding the piston for cutting the clip retaining grooves. (Piston is reversed for clarity)

The completed set of pistons for the 1930 Dodge is seen on the back cover of the book.

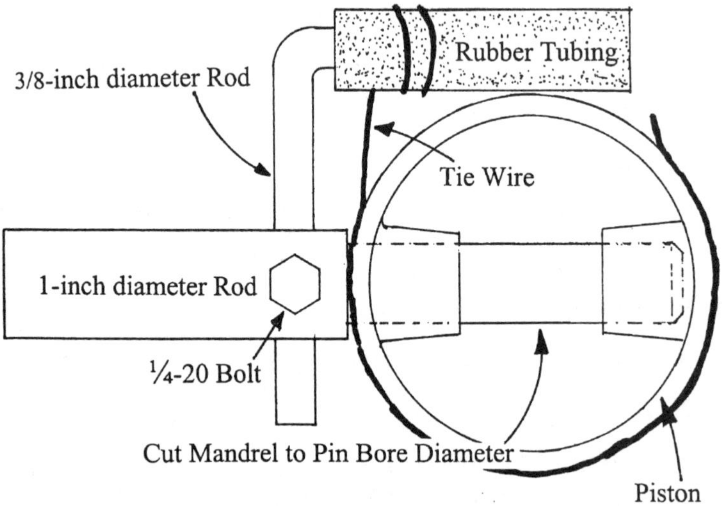

Pin-Bore Lathe Mandrel

CONCLUSION:

With a home shop and small foundry, you are able to bring engines to life for little cost. The 1930 Dodge pistons were made from scrap Chevy pistons that were melted in a furnace fired with used motor oil. The patterns were made from a few feet of yellow pine and some Bondo auto body filler. Sand, molasses and wheat paste were about $10. A complete Hastings ring set was about $80. The whole project requires approximately 7 to 10 days. In contrast, the one suppler I found wanted $1481.00 and 3 to 6 months for a set of pistons and rings. Clearly you can restore engines inexpensively. While the NASCAR teams might not be lining up at your door for a set of pistons, you can make parts that work and work well. It is quite a thrill to see your engine come to life with a healthy roar, especially when it is full of home made parts! I hope that you enjoy your projects as much as I enjoy mine.

Stephen D. Chastain

Bibliography

Blanchard, Harold F., *Motor's Auto Repair Manual* (New York: Hearst Magazines,1948)

Chastain, Stephen D., *Sand Casting Manual Vols. 1 &2,* (Jacksonville, Fl, 2004).

Heywood, John B., *Internal Combustion Engine Fundamentals* (New York: McGraw- Hill, 1988)

Howarth, M. H., *The Design of High Speed Diesel Engines,* (London: Constable, 1966).

Kent, William, *Kent's Mechanical Engineer's Handbook- Power,* (New York: John Wiley, 1936).

Obert, Edward F., *Internal Combustion Engines,*(Scranton: International Textbook Company, 1944).

Purday, H. F. P., *Diesel Engine Designing*, (Princeton, New Jersey: Van Nostrand, 1962).

Rohrle, Manfred D., *Pistons for Internal Combustion Engines,* (Germany: 1995).

Taylor, C.F., *The Internal Combustion Engine,*(Scranton, PA: International Textbook Company, 1961).

T.B.D. Service, *Ford Cars Anglia Perfect Popular Eight and Ten*, (London: A. Pearson Limited, 1934).

Trimble, George, S., *Design and Fabrication of Piston Rings*, (Strictly I.C., April 1989 Kent, WA).

Vallance, Alex, *Design of Machine Members,* (New York: McGraw-Hill, 1943).

INDEX: